THE
LITTLE PURPLE
PROBABILITY
BOOK

THE
LITTLE PURPLE
$\mathcal{P}$ROBABILITY
$\mathcal{B}$OOK

Master the Thinking Skills
to Succeed in Basic Probability

BRANDON ROYAL

M
MAVEN

Maven Publishing

Published by:

Maven Publishing
4520 Manilla Road
Calgary, Alberta, Canada T2G 4B7
www.mavenpublishing.com

Library and Archives Canada Cataloguing in Publication:

Royal, Brandon, author
The little purple probability book : master the thinking skills to succeed in basic probability / Brandon Royal.

Issued in print and electronic formats.

ISBN 978-1-897393-65-9 (paperback)
ISBN 978-1-897393-67-3 (ebook)

1. Probabilities. 2. Probabilities--Problems, exercises, etc. I. Title.

QA273.R68 2014   519.2   C2012-905731-2   C2012-905732-0

Cover Design: George Foster, Fairfield, Iowa, USA

This book's cover text was set in Minion.
The interior text was set in Scala and Scala Sans.

Contents

Introduction

This guidebook is compiled with one primary goal in mind: To help readers master basic probability in the shortest possible time frame.

Achieving this goal requires two steps. The first step is to develop a superior conceptual framework for understanding probability. The overview section at the beginning of this material contains two strategic flowcharts—one for probability and one for permutations and combinations—the two main topics under review. Included is a summary of all 14 essential formulas, each accompanied by a relevant example. The second step is to practice on a small number of the most classic, thematic problems that represent the broader types of probability problems that may be found elsewhere.

Mastering the thinking skills to succeed in basic probability requires an ability to match a given problem with the correct formula. In order to properly categorize a problem, it is necessary to understand the often confusing terminology that surrounds probability and its related topics. For example, what does it mean to say that two events are independent or that they are not mutually exclusive? Once we are confident that we can sort problems into their common types, we can then judge whether a particular problem is a variation based on a classic problem type and/or whether it contains common traps that are otherwise easy to overlook. Special notes (marked *NOTE* ⤴) are imbedded throughout this material. Here the reader will find additional information, examples, or follow-up problems that will no doubt prove useful.

This guidebook provides a complement to *The Little Green Math Book,* which addresses basic but tricky math involving arithmetic and algebra. Probability is a subject that warrants special attention and this book serves a stand-alone workshop.

Let's get started.

Overview

This material covers three main topics: probability, permutations, and combinations. What is the difference between probability and permutations and combinations? Probabilities are expressed as decimals or fractions between 0 and 1 (where 1 is the probability of certainty and 0 is the probability of impossibility) or, alternatively, as percentages between 0% and 100% (where 100% is the probability of certainty and 0% is the probability of impossibility). Permutations and combinations, on the other hand, result in outcomes greater than or equal to 1. Frequently they result in quite large outcomes such as 10, 36, 720, etc.

Probability:

In terms of probability, a quick rule of thumb is to determine first whether we are dealing with an "and" or "or" situation. "And" means multiply and "or" means addition. For example, if a problem states "what is the probability of x and y," we multiply individual probabilities together. If a problem states "what is the probability of x or y," we add individual probabilities together.

Moreover, if a probability problem requires us to multiply, we must ask one further question: "Are the events independent or are the events dependent?" *Independent* means that two events have no influence on one another and we simply multiply individual probabilities together to arrive at a final answer. *Not independent (dependent)* means that the occurrence of one event has an influence on the occurrence of another event and this influence must be taken into account.

Likewise, if a problem requires us to add probabilities, we must ask one further question: "Are the events mutually exclusive or non-mutually exclusive?" *Mutually exclusive* means that two events cannot occur at the same time and there is no "overlap" present. If two events have no overlap, we simply add probabilities. *Not mutually exclusive* means that two events can occur at the same time and that overlap is present. If two events do contain overlap, this overlap must not be double counted.

Permutations and Combinations:

With respect to permutations and combinations, permutations are ordered groups while combinations are unordered groups. Order matters in permutations; order does not matter in combinations. For example, AB and BA are considered different outcomes in permutations, but they are considered a single outcome in combinations. In real-life, examples of permutations include telephone numbers, license plates, electronic codes, and passwords. Examples of combinations include selection of members for a team or lottery tickets. In the case of lottery tickets, for instance, the order of numbers does not matter; we just need to get all the numbers, usually six of them.

As a practitioner's rule, the words "arrangements" or "possibilities" imply permutations; the words "select" or "choose" imply combinations.

Factorials:

Factorial means that we engage multiplication such that:

Example: $4! = 4 \times 3 \times 2 \times 1$
Example: $7! = 7 \times 6 \times 5 \times 4 \times 3 \times 2 \times 1$

Zero factorial equals one and one factorial also equals one:

Example: $0! = 1$
Example: $1! = 1$

Coins, Dice, Marbles, and Cards:

Problems in this guidebook include reference to coins, dice, marbles, and cards. For clarification purposes: The two sides of a coin are heads and tails. A die has six sides numbered from 1 to 6, with each having an equal likelihood of appearing subsequent to being tossed. The word "die" is singular; "dice" is plural. Marbles are assumed to be of a single, solid color. A deck of cards contains 52 cards divided equally into four suits—clubs, diamonds, hearts, and spades—where each suit contains 13 cards including ace, king, queen, jack, 10, 9, 8, 7, 6, 5, 4, 3, and 2.

EXHIBIT A   PROBABILITY FLOWCHART

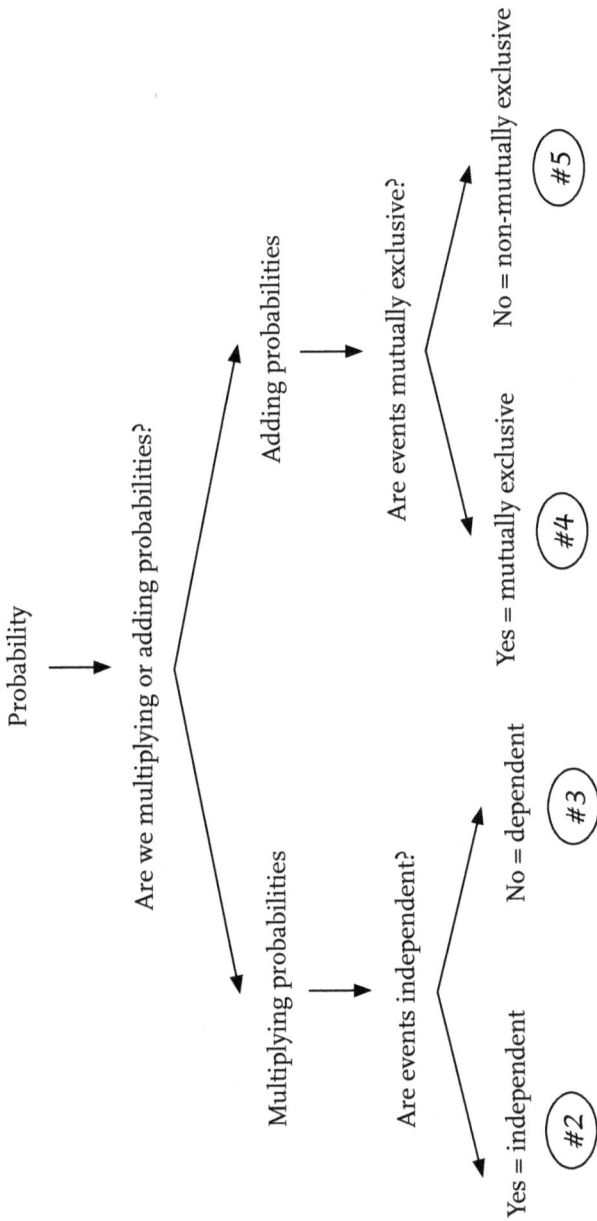

Probability

Are we multiplying or adding probabilities?

Multiplying probabilities

Adding probabilities

Are events independent?

Yes = independent

(#2)

No = dependent

(#3)

Are events mutually exclusive?

Yes = mutually exclusive

(#4)

No = non-mutually exclusive

(#5)

The numbers above denote the applicable probability formula.

EXHIBIT B   PERMUTATIONS AND COMBINATIONS FLOWCHART

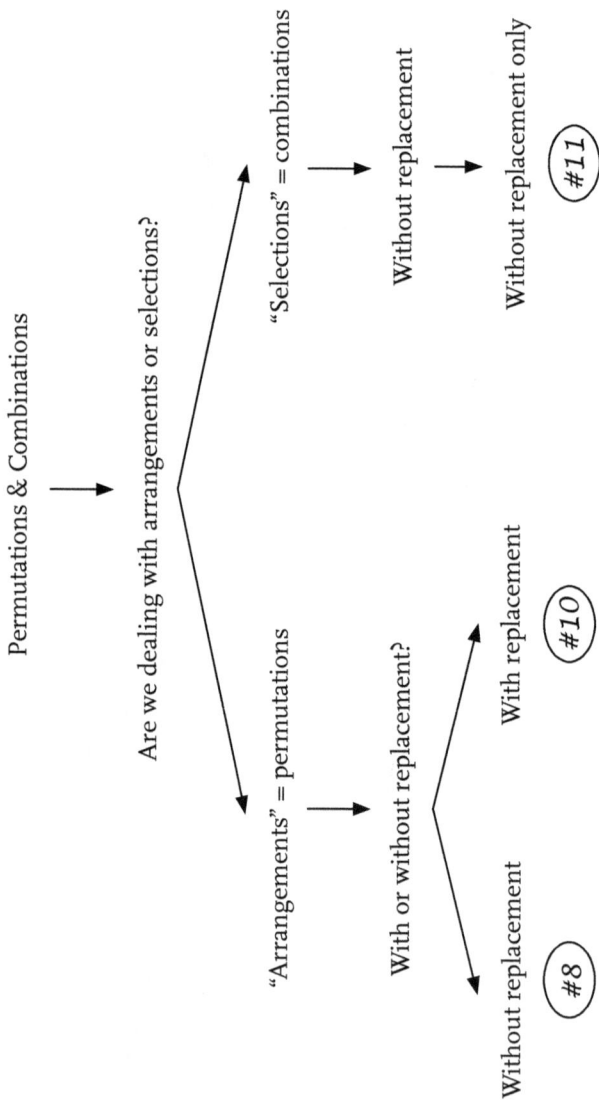

Permutations & Combinations

Are we dealing with arrangements or selections?

"Arrangements" = permutations

"Selections" = combinations

With or without replacement?

Without replacement

Without replacement   With replacement

(#8)   (#10)

Without replacement only

(#11)

The numbers above denote the applicable permutation or combination formula.

Basic Probability Formulas

Here at a glance are the basic probability, permutation, and combination formulas used in this material.

Universal Formula

#1 $$\text{Probability} = \frac{\text{Selected Events(s)}}{\text{Total Number of Possibilities}}$$

Example: You buy 3 raffle tickets and there are 10,000 tickets sold. What is the probability of winning the single prize?

$$\text{Probability} = \frac{3}{10,000}$$

Special Multiplication Rule

#2 $$P(A \text{ and } B) = P(A) \times P(B)$$

[Where the probability of A and B equals the probability of A times the probability of B]

If events are independent (that is, one event has no influence on the other), we simply multiply them together.

Example: What is the probability of tossing a coin twice and obtaining heads on both the first and second toss?

$$\frac{1}{2} \times \frac{1}{2} = \frac{1}{4}$$

General Multiplication Rule

(**#3**) $P(A\ and\ B) = P(A) \times P(B/A)$

[Where the probability of A and B equals the probability of A times the probability of B, given that A has already occurred]

If events are not independent (that is, one event has an influence on the other), we must adjust the second event based on its influence from the first event.

Example: A bag contains six marbles, three blue and three green. What is the probability of blindly reaching into the bag and pulling out two green marbles?

$$\frac{3}{6} \times \frac{2}{5} = \frac{6}{30} = \frac{1}{5}$$

Special Addition Rule

(**#4**) $P(A\ or\ B) = P(A) + P(B)$

[Where the probability of A or B equals the probability of A added to the probability of B]

If events are mutually exclusive (that is, there is no overlap), then we just add the probability of the events together.

Example: The probability that Sam will go to high school at Montcalm Academy is 50 percent, while the probability that he will go to Crescent Heights High School is 25 percent. What is the probability that he will choose to go to high school at either Montcalm Academy or Crescent Heights High School?

$50\% + 25\% = 75\%$

General Addition Rule

$$(\#5)\quad P(A \text{ or } B) = P(A) + P(B) - P(A \text{ and } B)$$

[Where the probability of A or B equals the probability of A added to the probability of B minus the probability of A and B]

If events are not mutually exclusive (that is, there is overlap), then we must subtract out the overlap subsequent to adding the events.

Example: The probability that tomorrow will be rainy is 30 percent. The probability that tomorrow will be windy is 20 percent. What is the probability that tomorrow's weather will be either rainy or windy?

30% + 20% − (30% × 20%)
30% + 20% − 6%
50% − 6% = 44%

With regard to the General Addition Rule, the reason that we subtract out the overlap is because we do not want to count it twice. When two events overlap, both events contain that same overlap. Thus, it must be subtracted once in order not to "double" count it.

NOTE ∾ Let's quickly contrast what is commonly referred to as the inclusive "or" and the exclusive "or." The problem used in support of probability formula #5 is governed by an inclusive "or." It is reasonable to assume that tomorrow's weather can be <u>both</u> rainy and windy. The problem is effectively asking: "What is the probability that tomorrow's weather will be either rainy, or windy, or both rainy and windy?" The inclusive "or" occurs whenever there is overlap. The problem used in support of probability formula #4 effectively asks: "What is the probability that Sam will choose to go to high school at Montcalm Academy or

Crescent Heights High School, but not both high schools?" The choice between going to high school at two different locations is a mutually exclusive one, and we treat that particular problem as involving an exclusive "or."

Complement Rule

#6 $P(A) = 1 - P(\text{not } A)$

[Where the probability of A equals one minus the probability of A not occurring]

The Complement Rule of probability describes the *subtracting* of probabilities rather than the adding or multiplying of probabilities. To calculate the probability of an event using this rule, we ask: "What is the probability of a given event not occurring?" Then we subtract this result from 1.

Example: What is the probability of rolling a pair of dice and not rolling double sixes?

The probability of rolling double sixes:

$$\frac{1}{6} \times \frac{1}{6} = \frac{1}{36}$$

The probability of not rolling double sixes:

$$1 - \frac{1}{36} = \frac{35}{36}$$

Rule of Enumeration

#7 If there are x ways of doing one thing, y ways of doing a second thing, and z ways of doing a third thing, then the number of ways doing all these things is $x \times y \times z$. This is known as the Rule of Enumeration.

NOTE ∽ Technically, the Rule of Enumeration does not fall under the umbrella of "probability" or "permutation" or "combination." For practical reasons, however, it is most often discussed along with probability.

Example: Fast-Feast Restaurant offers customers a set menu with a choice of one of each of the following: 2 different salads, 3 different soups, 5 different entrees, 3 different desserts, and coffee or tea. How many possibilities are there with respect to how a customer can take his or her meal?

$$2 \times 3 \times 5 \times 3 \times 2 = 180$$

Permutations

(#8) Without replacement $\qquad _nP_r = \dfrac{n!}{(n-r)!}$

[Where n = total number of items and r = number of items we are taking or arranging]

Example: How many two-letter codes can be made from the letters A, B, C, and D if the same letter cannot be displayed more than once in any given code?

$$_nP_r = \dfrac{n!}{(n-r)!}$$

$$_4P_2 = \dfrac{4!}{(4-2)!} = \dfrac{4!}{2!} = \dfrac{4 \times 3 \times \cancel{2 \times 1}}{\cancel{2 \times 1}} = 4 \times 3 = 12$$

(#9) $_nP_n = n!$

[Shortcut formula when all items are taken together]

Example: How many ways can a person display (or arrange) four different books on a shelf?

$$_nP_r = \dfrac{n!}{(n-r)!}$$

$$_4P_4 = \frac{4!}{(4-4)!} = \frac{4!}{0!} = \frac{4!}{1} = 4! = 4 \times 3 \times 2 \times 1 = 24$$

Also, shortcut formula: $n! = 4! = 24$

(#10) With replacement n^r

Example: How many four-digit codes can be made from the numbers 1, 2, 3, and 4, if the same numbers can be displayed more than once in any given code?

$$n^r \qquad\qquad 4^4 = 256$$

NOTE ⤚ Permutation with replacement (that is, n^r) technically falls under the Rule of Enumeration. It is included here for ease of presentation. For a problem to be considered a permutation, the permutation formula must be applicable.

Combinations

(#11) $_nC_r = \dfrac{n!}{r!(n-r)!}$

[Where n = total number of items taken and r = the number of items we are choosing or selecting]

Example: How many ways can a person choose three of four colors for the purpose of painting the inside of a house?

$$_nC_r = \frac{n!}{r!(n-r)!}$$

$$_4C_3 = \frac{4!}{3!(4-3)!} = \frac{4!}{3! \times (1)!} = \frac{4 \times \cancel{3 \times 2 \times 1}}{\cancel{3 \times 2 \times 1} \times 1} = 4$$

Additional formulas:

Joint Permutations

$$\text{(#12)} \quad _nP_r \times {}_nP_r = \frac{n!}{(n-r)!} \times \frac{n!}{(n-r)!}$$

Example: A tourist plans to visit three of five Western European cities and then proceed to visit two of four Eastern European cities. At the planning stage, how many itineraries are possible?

$$_5P_3 \times {}_4P_2$$

$$\frac{5!}{(5-3)!} \times \frac{4!}{(4-2)!}$$

$$\frac{5!}{2!} \times \frac{4!}{2!}$$

$$\frac{5 \times 4 \times 3 \times \cancel{2 \times 1}}{\cancel{2 \times 1}} \times \frac{4 \times 3 \times \cancel{2 \times 1}}{\cancel{2 \times 1}}$$

$$60 \times 12 = 720$$

Multiplying outcomes, rather than adding them, is consistent with the treatment afforded by the Rule of Enumeration.

Joint Combinations

$$\text{(#13)} \quad _nC_r \times {}_nC_r = \frac{n!}{r!(n-r)!} \times \frac{n!}{r!(n-r)!}$$

Example: A special marketing task force is to be chosen from five professional golfers and five professional tennis players. If the final task force chosen is to consist of three golfers and three tennis players, then how many different task forces are possible?

$$_5C_3 \times {}_5C_3$$

$$\frac{5!}{3!\,(5-3)!} \times \frac{5!}{3!\,(5-3)!}$$

$$\frac{5!}{3!\,(2)!} \times \frac{5!}{3!\,(2)!}$$

$$\frac{5 \times 4 \times \cancel{3 \times 2 \times 1}}{\cancel{3 \times 2 \times 1} \times 2 \times 1} \times \frac{5 \times 4 \times \cancel{3 \times 2 \times 1}}{\cancel{3 \times 2 \times 1} \times 2 \times 1}$$

$$10 \times 10 = 100$$

Repeated Letters or Numbers (Permutations)

#14 $\dfrac{n!}{x!\,y!\,z!}$ [where x, y, and z are different but identical letters or numbers]

Example: How many four-numeral codes can be created using the four numbers 0, 0, 1, and 2?

$$\frac{4!}{2!} = \frac{4 \times 3 \times \cancel{2 \times 1}}{\cancel{2 \times 1}} = 12$$

Note that 2! denotes the two zeros which represent repeated numbers.

Probability

1. **Four Aces**

   Which of the following represents the probability of selecting four cards at random from a deck of cards and getting four aces? (The cards are to be selected one after the other without replacing any of the cards.)

   A) $\dfrac{1}{52} \times \dfrac{1}{52} \times \dfrac{1}{52} \times \dfrac{1}{52}$

   B) $\dfrac{1}{52} \times \dfrac{1}{51} \times \dfrac{1}{50} \times \dfrac{1}{49}$

   C) $\dfrac{4}{52} \times \dfrac{3}{51} \times \dfrac{2}{50} \times \dfrac{1}{49}$

   D) $\dfrac{4}{52} \times \dfrac{3}{52} \times \dfrac{2}{52} \times \dfrac{1}{52}$

   E) $\dfrac{4}{52} \times \dfrac{4}{52} \times \dfrac{4}{52} \times \dfrac{4}{52}$

2. **Orange & Blue**

   There are 5 marbles in a bag—2 are orange and 3 are blue. If two marbles are pulled from the bag, what is the probability that the first will be orange and the second will be blue?

   A) $\dfrac{6}{25}$ B) $\dfrac{3}{10}$ C) $\dfrac{2}{5}$ D) $\dfrac{3}{5}$ E) $\dfrac{7}{10}$

3. Orange & Blue Again

There are 5 marbles in a bag—2 are orange and 3 are blue. If two marbles are pulled from the bag, what is the probability that at least one will be orange?

A) $\dfrac{6}{25}$ B) $\dfrac{3}{10}$ C) $\dfrac{2}{5}$ D) $\dfrac{3}{5}$ E) $\dfrac{7}{10}$

4. Exam Time

A student is to take her final exams in two subjects. The probability that she will pass the first subject is $\dfrac{3}{4}$ and the probability that she will pass the second subject is $\dfrac{2}{3}$. What is the probability that she will pass one exam or the other exam?

A) $\dfrac{5}{12}$ B) $\dfrac{1}{2}$ C) $\dfrac{7}{12}$ D) $\dfrac{5}{7}$ E) $\dfrac{11}{12}$

5. Sixth Sense

What is the probability of rolling a six on either the first or second toss of a dice?

A) $\dfrac{1}{36}$ B) $\dfrac{5}{18}$ C) $\dfrac{1}{6}$ D) $\dfrac{11}{36}$ E) $\dfrac{1}{3}$

6. Exam Time Encore

A student is to take her final exams in three subjects. The probability that she will pass the first subject is $\frac{3}{4}$, the probability that she will pass the second subject is $\frac{2}{3}$, and the probability that she will pass the third subject is $\frac{1}{2}$. What is the probability that she will pass at least one of these three exams? (Or: What is the probability that she will pass either the first exam or the second exam or the third exam?)

A) $\frac{1}{4}$ B) $\frac{11}{24}$ C) $\frac{17}{24}$ D) $\frac{3}{4}$ E) $\frac{23}{24}$

Enumerations

7. Hiring

A company seeks to hire a sales manager, a shipping clerk, and a receptionist. The company has narrowed its candidate search and plans to interview all remaining candidates including 7 persons for the position of sales manager, 4 persons for the position of shipping clerk, and 10 persons for the position of receptionist. How many different hirings of these three people are possible?

A) $7 \times 4 \times 10$

B) $7 + 4 + 10$

C) $21 \times 20 \times 19$

D) $7! + 4! + 10!$

E) $7! \times 4! \times 10!$

Permutations

8. Fencing

Four contestants representing four different countries advance to the finals of a fencing championship. Assuming all competitors have an equal chance of winning, how many possibilities are there with respect to how a first-place and second-place medal can be awarded?

A) 6
B) 7
C) 12
D) 16
E) 24

9. Row

Six students are to sit in a row side by side for a makeup exam. How many ways could they arrange themselves?

A) 12
B) 36
C) 72
D) 240
E) 720

10. Alternating

Six students—3 boys and 3 girls—are to sit in a row side by side for a makeup exam. How many ways could they arrange themselves given that no two boys and no two girls can sit next to one another?

A) 12
B) 36
C) 72
D) 240
E) 720

11. Banana

Which of the following leads to the correct mathematical solution for the number of ways that the letters of the word BANANA could be arranged to create a six-letter code?

A) $6!$

B) $6! - (3! \times 2!)$

C) $6! - (3! + 2!)$

D) $\dfrac{6!}{3! \times 2!}$

E) $\dfrac{6!}{3! + 2!}$

12. Table

How many ways could three people sit at a table with five seats in which two of the five seats will remain empty?

A) 8
B) 12
C) 60
D) 118
E) 120

Combinations

13. Singer

For an upcoming charity event, a male vocalist has agreed to sing 4 out of 6 "old songs" and 2 out of 5 "new songs." How many ways can the singer make his selection?

A) 25
B) 50
C) 150
D) 480
E) 600

14. Reunion

If 11 people meet at a reunion and each person shakes hands exactly once with each of the others, what is the total number of handshakes?

A) $11 \times 10 \times 9 \times 8 \times 7 \times 6 \times 5 \times 4 \times 3 \times 2 \times 1$
B) $10 \times 9 \times 8 \times 7 \times 6 \times 5 \times 4 \times 3 \times 2 \times 1$
C) 11×10
D) 55
E) 45

15. Outcomes

Given that $_nP_r = \dfrac{n!}{(n-r)!}$ and $_nC_r = \dfrac{n!}{r!\,(n-r)!}$, where n is the total number of items and r is the number of items taken or chosen, which of the following statements are true in terms of the number of outcomes generated?

I. $_5P_3 > {_5P_2}$

II. $_5C_3 > {_5C_2}$

III. $_5C_2 > {_5P_2}$

A) I only
B) I & II only
C) I & III only
D) II & III only
E) I, II & III

Answers and Explanations

1. Four Aces

Choice C
Overview: This problem presents a classic application of the General Multiplication Rule of probability. Selecting a card "without replacement" affects or influences the probability of the next card being chosen because the first card is missing from the deck. Selecting a card "with replacement" does not affect or influence the probability of the next card being chosen because replacing a given card restores the deck to its prior state.

On the first pick, there is a $\frac{4}{52}$ chance of selecting an ace. On our second pick, there is a $\frac{3}{51}$ chance of selecting an ace because there is one fewer ace to choose from and one fewer card in the deck. On the third pick, there is a $\frac{2}{50}$ chance of selecting an ace because there is one fewer ace to choose from and one fewer card in the deck. Finally, there is a $\frac{1}{49}$ chance of selecting an ace because there is only one ace to choose from with exactly 49 cards left in the deck.

For the record, answer choice E would have been the correct answer if we had selected our four aces with replacement. Such would be the case if the problem had asked: "What is the probability of selected four aces in a row if we replace each card before selecting the next one?" Choice A would be the correct answer if we were to choose any single card with replacement. For example: "What is the probability of selecting the ace of spades four times in a row from a random deck of cards, if we replace that card in the deck after selecting it?" Choice B would have been the correct answer if we had to select these four aces in a specific order and had done so without replacing each card. Such would be the case if the problem had asked: "What is the probability of selecting the ace of spades, followed by the ace of hearts, followed by the ace of clubs, and followed by the ace of diamonds, if we do not replace each card after selecting it?"

NOTE ❧ There are two sets of terms that are key to understanding how to distinguish among basic probability problems. The first set of terms is "mutually exclusive" and "not mutually exclusive." The second set of terms is "independent" and "not independent (dependent)." *Mutually exclusive* means that two events or outcomes do not overlap with one another or cannot occur at the same time. *Not mutually exclusive* means that two events or outcomes do overlap with one another or can occur at the same time. *Independent* means that two events or outcomes do not influence one another and occur randomly relative to each other. *Not independent (or dependent)* means that two events or outcomes influence one another and that the occurrence of one event has an affect on the occurrence of another event.

Here are some simple real-life examples to illuminate these terms. Say we are putting on a business conference and inviting attendees as well as guest speakers. The assignment of VIP seating and non-VIP seating is a mutually exclusive outcome. Either a person is in the VIP seats or he or she is not. The same holds true for determining who is an in-state versus out-of-state attendee. An attendee is either in-state or out-of-state with no overlap possible. However, in classifying attendees by profession, we might have overlap between who is a manager and who is an engineer and who is a salesperson and who is an entrepreneur. Obviously, some attendees might fall into more than one category. These categories would, therefore, not be mutually exclusive.

Two tasks might be independent and have no influence on one another. For example, in preparing for the conference, it wouldn't make any difference whether we made name tags first and then made copies of the conference handouts or made copies of the conference handouts and then made name tags. These events represent separate tasks that have no bearing on one another. On the other hand, two tasks may not be independent of one another;

they may, in fact, be dependent on one another. This is true of events that must occur in a certain sequence. In preparing for the conference, we must plan the conference first before inviting speakers to speak at the conference. Likewise, attendees must be registered for the conference, before they can be admitted to the conference and before they ever fill out conference evaluation forms, which are handed out at the end of the conference. In other words, the filling out of a conference evaluation form is dependent upon a person actually attending the conference, which, in turn, is dependent upon a person having first registered for the conference.

2.   Orange & Blue

Choice B
Overview: This problem is solved using the General Multiplication Rule. The use of the word "and" in this problem is a signal that we need to multiply probabilities, not add probabilities. The probability of choosing the first orange marble will influence the probability of choosing the first blue marble because there will be one fewer marble to choose from (selection without replacement).

The probability of first choosing an orange marble is $\frac{2}{5}$. The probability of then choosing a blue marble is $\frac{3}{4}$.

$$\frac{2}{5} \times \frac{3}{4} = \frac{6}{20} = \frac{3}{10}$$

Note that the trap answer, choice A, involves forgetting to subtract one marble from the denominator of the second fraction:

$$\frac{2}{5} \times \frac{3}{5} = \frac{6}{25}$$

3. Orange & Blue Again

Choice E

Overview: This problem introduces the Complement Rule of probability. See probability formula #6 (page 16).

The simplest way to view this problem is in terms of what we don't want. At least one orange marble means that we want anything but two blue marbles.

Probability of getting two blue marbles:

$$\frac{3}{5} \times \frac{2}{4} = \frac{6}{20} = \frac{3}{10}$$

Therefore, the probability of getting at least one orange marble is the same as one minus the probability of getting two blue marbles.

$$P(A) = 1 - P(\text{not } A)$$

$$1 - \frac{3}{10} = \frac{7}{10}$$

Another way to solve this problem is using the direct method, which entails writing out all possibilities and adding up the results that we are looking for. There are four possible outcomes when we choose two marbles at random. Three of these outcomes yield at least one orange marble:

Orange, Blue:	$\dfrac{2}{5} \times \dfrac{3}{4} = \dfrac{6}{20}$
Blue, Orange:	$\dfrac{3}{5} \times \dfrac{2}{4} = \dfrac{6}{20}$
Orange, Orange:	$\dfrac{2}{5} \times \dfrac{1}{4} = \dfrac{2}{20}$
Blue, Blue:	$\dfrac{3}{5} \times \dfrac{2}{4} = \dfrac{6}{20}$

$\left. \right\} \quad \dfrac{14}{20} \Rightarrow \dfrac{7}{10}$

Note that the total of all of the above possibilities equals 1 (that is, $\dfrac{6}{20} + \dfrac{6}{20} + \dfrac{2}{20} + \dfrac{6}{20} = \dfrac{20}{20} = 1$) because there are no other possibilities other than the four outcomes presented here.

4. Exam Time

Choice E
Overview: The use of the word "or" in this problem (that is, "pass one exam or the other exam") is a signal that probabilities are to be added, not multiplied. The probability of two non-mutually exclusive events A or B is calculated by adding the probability of the first event to the second event and then subtracting out the overlap between the two events. This is referred to in probability as the General Addition Rule. See probability formula #5 (page 15).

$$P(A \text{ or } B) = P(A) + P(B) - P(A \text{ and } B)$$

$$\frac{9}{12} + \frac{8}{12} - \frac{6}{12} = \frac{11}{12}$$

Here, the probability of passing the first exam is added to the probability of passing the second exam, less the probability of passing both exams. The probability of passing both exams is calculated as follows: $\dfrac{3}{4} \times \dfrac{2}{3} = \dfrac{6}{12} = \dfrac{1}{2}$. If we don't make this subtraction, we will double count the possibility that she will pass both exams.

NOTE One way to prove this result is to recognize that the probability of passing either exam (including passing both exams) is everything other than failing both exams. The probability of failing both exams is $\frac{1}{4} \times \frac{1}{3} = \frac{1}{12}$. Therefore, the probability of passing either the first exam or the second exam (or both exams) is $1 - \frac{1}{12} = \frac{11}{12}$.

5. Sixth Sense

Choice D
Overview: The use of the word "or" signals the need to add probabilities (that is, "first or second toss of a dice"). The only other question to be asked is whether there is a need to subtract out overlap. That is, "Are the two events mutually exclusive?" As it turns out, they are not mutually exclusive and this overlap must be subtracted out.

Let's use a chart to visualize all the possibilities that occur when a dice is rolled twice.

Second Roll

		1	2	3	4	5	6
	1	1,1	1,2	1,3	1,4	1,5	**1,6**
	2	2,1	2,2	2,3	2,4	2,5	**2,6**
First Roll	3	3,1	3,2	3,3	3,4	3.5	**3,6**
	4	4,1	4,2	4,3	4,4	4,5	**4,6**
	5	5,1	5,2	5,3	5,4	5,5	**5,6**
	6	**6,1**	**6,2**	**6,3**	**6,4**	**6,5**	6,6

As seen in the chart above, there are of course thirty-six possible outcomes when we roll a single dice twice (or, equally, if we toss a pair of dice). As indicated by the bolded numbers in the chart, we have 11 outcomes with respect to how we can get exactly one six: (6,1), (6,2), (6,3), (6,4), (6,5), (6,6) and (1,6), (2,6), (3,6), (4,6), (5,6).

The applicable formula for solving this particular problem is the General Addition Rule of probability:

$$P(A \text{ or } B) = P(A) + P(B) - P(A \text{ and } B)$$

$$\frac{1}{6} + \frac{1}{6} - \frac{1}{36}$$

$$\frac{6}{36} + \frac{6}{36} - \frac{1}{36}$$

$$\frac{12}{36} - \frac{1}{36} = \frac{11}{36}$$

Note that we can't just add $\frac{1}{6}$ and $\frac{1}{6}$ to get $\frac{12}{36}$ or $\frac{1}{3}$, which incidentally is answer choice E. To do so would fail to account for, and properly remove, the overlap created when double sixes are rolled.

After all, what is the probability of getting a six on the first roll of a dice? Answer: (6,1), (6,2), (6,3), (6,4), (6,5), and (6,6). The probability of event A is $\frac{6}{36}$ or $\frac{1}{6}$. What is the probability of getting a six on the second roll of a dice? Answer: (1,6), (2,6), (3,6), (4,6), (5,6), and (6,6). The probability of event B is $\frac{6}{36}$ or $\frac{1}{6}$. Note that "double sixes" is included in both event A and event B. This overlap must be subtracted out. To be clear, the probability of getting a six on the first or second roll of a dice does include the possibility of getting sixes on the first and second rolls (that is, double sixes), but this outcome can only be counted once.

Another way this problem could have been asked is: "What is the probability of rolling two normal six-sided dice and getting at least one six?" And yet another way this problem could have been solved is through the use of the Complement Rule of probability. The probability of rolling at least one six is the same as the probability of "one minus the probability of rolling no sixes."

The probability of rolling no sixes is:

$$\frac{5}{6} \times \frac{5}{6} = \frac{25}{36}$$

The probability of rolling at least one six is:

$$1 - \frac{25}{36} = \frac{11}{36}$$

NOTE ✎ Answer choice B would have been the correct answer had the problem asked: "What is the probability of rolling a single dice twice and getting <u>exactly</u> one six?" Perhaps the simplest way to arrive at the correct answer is to write out the possibilities. There are eleven ways to get exactly one six: (6,1), (6,2), (6,3), (6,4), (6,5), (6,6), (1,6), (2,6), (3,6), (4,6), (5,6), and (6,6). Alternatively, we could choose to subtract out $\frac{1}{36}$ from the previous calculation (i.e., $\frac{11}{36}$) in order to remove the probability of rolling double sixes.

Note that in the following calculation, the "first" six is removed because it represents overlap while the "second" six is removed because it represents the actual probability of rolling double sixes.

$$\frac{1}{6} + \frac{1}{6} - \frac{1}{36} - \frac{1}{36} = \frac{10}{36} = \frac{5}{18}$$

6. Exam Time Encore

Choice E
Overview: This problem involves three overlapping probabilities and, as evident by the "shortcut" approach, it is best solved using the Complement Rule of probability.

I. Shortcut Approach

Using the Complement Rule, we want to calculate the probability of failing all three exams. Then we will subtract this number from 1, in order to determine the probability of passing any and all exams.

i) The probability of <u>not</u> passing the first exam:

$$P(\text{not } A) = 1 - P(A) \qquad 1 - \frac{3}{4} = \frac{1}{4}$$

ii) Below is the probability of <u>not</u> passing the second exam:

$$P(\text{not } B) = 1 - P(B) \qquad 1 - \frac{2}{3} = \frac{1}{3}$$

iii) Below is the probability of <u>not</u> passing the third exam:

$$P(\text{not } C) = 1 - P(C) \qquad 1 - \frac{1}{2} = \frac{1}{2}$$

iv) The probability of <u>failing all three</u> exams:

$$P(\text{not } A \text{ or } B \text{ or } C) \qquad \frac{1}{4} \times \frac{1}{3} \times \frac{1}{2} = \frac{1}{24}$$

v) The probability of <u>passing at least one</u> exam:

$$P(A) = 1 - P(\text{not } A) \qquad 1 - \frac{1}{24} = \frac{23}{24}$$

II. Direct Approach

Using the direct approach, we calculate the probability of passing only one of the three exams, two of the three exams, and all of the three exams. Then, we add these seven results together.

1. Probability of <u>passing exam one</u> but not exams two or three:

$$P(A) \times P(\text{not } B) \times P(\text{not } C) \qquad \frac{3}{4} \times \frac{1}{3} \times \frac{1}{2} = \frac{3}{24}$$

2. Probability of <u>passing exam two</u> but not exams one or three:

$P(\text{not } A) \times P(B) \times P(\text{not } C)$ $\qquad \dfrac{1}{4} \times \dfrac{2}{3} \times \dfrac{1}{2} = \dfrac{2}{24}$

3. Probability of <u>passing exam three</u> but not exams one or two:

$P(\text{not } A) \times P(\text{not } B) \times P(C)$ $\qquad \dfrac{1}{4} \times \dfrac{1}{3} \times \dfrac{1}{2} = \dfrac{1}{24}$

4. Probability of <u>passing exams one</u> and <u>two</u> but not exam three:

$P(A) \times P(B) \times P(\text{not } C)$ $\qquad \dfrac{3}{4} \times \dfrac{2}{3} \times \dfrac{1}{2} = \dfrac{6}{24}$

5. Probability of <u>passing exams one</u> and <u>three</u> but not exam two:

$P(A) \times P(\text{not } B) \times P(C)$ $\qquad \dfrac{3}{4} \times \dfrac{1}{3} \times \dfrac{1}{2} = \dfrac{3}{24}$

6. Probability of <u>passing exams two</u> and <u>three</u> but not exam one:

$P(\text{not } A) \times P(B) \times P(C)$ $\qquad \dfrac{1}{4} \times \dfrac{2}{3} \times \dfrac{1}{2} = \dfrac{2}{24}$

7. Probability of <u>passing all three</u> exams:

$P(A) \times P(B) \times P(C)$ $\qquad \dfrac{3}{4} \times \dfrac{2}{3} \times \dfrac{1}{2} = \dfrac{6}{24}$

8. Probability of <u>not passing any</u> of the three exams:

$P(\text{not } A) \times P(\text{not } B) \times P(\text{not } C)$ $\qquad \dfrac{1}{4} \times \dfrac{1}{3} \times \dfrac{1}{2} = \dfrac{1}{24}$

These are all the possibilities regarding the outcomes of one student taking three exams. Adding the first seven of eight possibilities above will result in the correct answer using the direct approach.

Proof:
$$\frac{3}{24} + \frac{2}{24} + \frac{1}{24} + \frac{6}{24} + \frac{3}{24} + \frac{2}{24} + \frac{6}{24} = \frac{23}{24}$$

Note that the total of all eight outcomes above will total to 1 because 1 is the sum total of all probabilistic possibilities.

Proof:
$$\frac{3}{24} + \frac{2}{24} + \frac{1}{24} + \frac{6}{24} + \frac{3}{24} + \frac{2}{24} + \frac{6}{24} + \frac{1}{24} = \frac{24}{24} = 1$$

7. Hiring

Choice A
Overview: This particular problem is frequently mistaken for a permutation problem, but does not fall under the umbrella of probability or permutation or combination.

The solution requires only that we multiply together all individual possibilities. Multiplying 7 (that is, candidates for sales managers) by 4 (that is, candidates for shipping clerk) by 10 (that is, candidates for receptionist) would result in 280 possibilities.

$$7 \times 4 \times 10 = 280$$

NOTE ৯ This problem is about a series of independent choices. It utilizes the "multiplier principle" and falls within the Rule of Enumeration. The permutation formula is not applicable and cannot be used with this type of problem. This problem is concerned with *how many options we have,* not *how many arrangements are possible,* as is the case when dealing with permutation problems.

8. Fencing

Choice C
Overview: This problem is a permutation problem, not a combination problem. In permutation problems, order matters. If country A wins the tournament and country B places second, it is a different outcome than if country B wins and country A places second.

$$_nP_r = \frac{n!}{(n-r)!}$$

$$_4P_2 = \frac{4!}{(4-2)!} = \frac{4!}{(2)!} = \frac{4 \times 3 \times \cancel{2 \times 1}}{\cancel{2 \times 1}} = 12$$

NOTE ✍ Consider this follow-up problem. A teacher has four students in a special needs class. She must assign four awards at the end of the year—math, English, history, and creative writing awards. How many ways could she do this assuming that a single student could win multiple awards?

$$n^r = 4^4 \qquad 4 \times 4 \times 4 \times 4 = 256$$

She has four ways that she could give out the math award, four ways to give out the English award, four ways to give out the history award, and four ways to give out the creative writing award. Refer to probability formula #10 (page 18).

9. Row

Choice E
Overview: This is a permutation problem which lends itself to the shortcut formula—$n!$—given that we are utilizing all members of set n, as opposed to a subset of set n.

The problem is essentially asking us: "How many ways can we arrange six people in six seats?" The fact that three of these six individuals are boys and three are girls is irrelevant to the problem at hand. If there are no restrictions on how the students may be seated, there are 720 possibilities with respect to how they can be seated.

$$6! = 6 \times 5 \times 4 \times 3 \times 2 \times 1 = 720 \text{ possibilities}$$

There are six ways to seat the first student, five ways to seat the second student, four ways to seat the third student, three ways to seat the fourth student, two ways to seat the fifth student, and only one way to seat the sixth and final student.

10. Alternating

Choice C

Overview: This problem is effectively a joint permutation problem in which we calculate two individual permutations and multiply those outcomes together.

There are two possibilities with respect to how the girls and boys can sit for the make-up exam. Per scenario 1, a boy will sit in the first, third, and fifth seats and a girl will sit in the second, fourth, and sixth seats. Alternatively, per scenario 2, a girl will sit in the first, third, and fifth seats and a boy will sit in the second, fourth, and sixth seats.

Scenario 1:

$$\begin{array}{cccccc} B & G & B & G & B & G \end{array}$$

$$\frac{3}{B_1} \times \frac{3}{G_1} \times \frac{2}{B_2} \times \frac{2}{G_2} \times \frac{1}{B_3} \times \frac{1}{G_3}$$

Scenario 2:

$$\begin{array}{cccccc} G & B & G & B & G & B \end{array}$$

$$\frac{3}{G_1} \times \frac{3}{B_1} \times \frac{2}{G_2} \times \frac{2}{B_2} \times \frac{1}{G_3} \times \frac{1}{B_3}$$

With reference to scenario 1, how many ways can each seat be filled (left to right)? Answer: The *first* seat can be filled by one of three boys, the *second* seat can be filled by one of three girls, the *third* seat can filled by one of two remaining boys, the *fourth* seat can be filled by one of two remaining girls, the *fifth* seat will be filled by the final boy, and the *sixth* seat will be filled by the final girl.

With reference to scenario 2, how many ways can each seat be filled (left to right)? Answer: The *first* seat can be filled by one of three girls, the *second* seat can be filled by one of three boys, the *third* seat can filled by one of two remaining girls, the *fourth* seat can be filled by one of two remaining boys, the *fifth* seat will be filled by the final girl, and the *sixth* seat will be filled by the final boy.

Therefore:

B G B G   B G   G B G   B G B

$(3 \times 3 \times 2 \times 2 \times 1 \times 1) + (3 \times 3 \times 2 \times 2 \times 1 \times 1)$
$36 + 36 = 72$

In short, the answer can be calculated as follows:

$(3! \times 3!) + (3! \times 3!)$
$2(3! \times 3!)$
$2[(3 \times 2 \times 1) \times (3 \times 2 \times 1)] = 72$
$2(6 \times 6) = 72$
$2(36) = 72$

NOTE ✑ There is another common variation stemming from this type of permutation problem:

> Three boys and three girls are going to sit for a make-up exam. The girls are to sit in the first, second, and third seats while the boys must sit in the fourth, fifth, and sixth seats. How many possibilities are there with respect to how the six students can be seated?

$$\underset{B_1}{\overset{B}{\frac{3}{B_1}}} \times \underset{G_1}{\overset{G}{\frac{3}{G_1}}} \times \underset{B_2}{\overset{B}{\frac{2}{B_2}}} \times \underset{G_2}{\overset{G}{\frac{2}{G_2}}} \times \underset{B_3}{\overset{B}{\frac{1}{B_3}}} \times \underset{G_3}{\overset{G}{\frac{1}{G_3}}}$$

Answer: $3! \times 3! = 6 \times 6 = 36$ possibilities

11. Banana

Choice D

Overview: This problem highlights the handling of "repeated letters" (or "repeated numbers"). The formula for calculating permutations with repeated numbers or letters is $\frac{n!}{x!\,y!\,z!}$, where x, y, and z are distinct but identical numbers or letters.

$$\frac{n!}{x!\,y!} = \frac{6!}{3! \times 2!} = \frac{6 \times 5 \times 4 \times \cancel{3 \times 2 \times 1}}{(\cancel{3 \times 2 \times 1}) \times (2 \times 1)} = 60$$

The word "banana" has three "a's" and two "n's"—3! denotes the three "a's" while 2! denotes the two "n's."

12. Table

Choice C

Overview: This problem deals with the prickly issue of "empty seats."

$$\frac{5!}{2!} = \frac{5 \times 4 \times 3 \times \cancel{2 \times 1}}{\cancel{2 \times 1}} = 60$$

2! represents the two empty seats.

NOTE ✐ The answer to this problem is similar in approach to that of the problem 11, *Banana*. In permutation theory, "empty seats" are analogous to "identical numbers" (or "identical letters"). Think of the two empty chairs as representing two identical people. Also, the geometric configuration of a round table should not be a distraction. The solution to this problem would be identical had we been dealing with a row of five seats. After all, a table is but a row attached at both its ends and, in the case of a round table, shaped like a circle.

13. Singer

Choice C
Overview: Joint combination problems are solved by multiplying the results of two individual combinations.

First, break the combination into two calculations. First, the "old songs," and second, the "new songs."

Old songs:

$$_nC_r = \frac{n!}{r!(n-r)!}$$

$$_6C_4 = \frac{6!}{4!(6-4)!} = \frac{6!}{4!(2)!} = \frac{6 \times 5 \times \cancel{4 \times 3 \times 2 \times 1}}{\cancel{4 \times 3 \times 2 \times 1} \times (2 \times 1)} = 15$$

Thus, 15 represents the number of ways the singer can choose to sing four of six old songs.

New songs:

$$_nC_r = \frac{n!}{r!(n-r)!}$$

$$_5C_2 = \frac{5!}{2!(5-2)!} = \frac{5!}{2!(3)!} = \frac{5 \times 4 \times \cancel{3 \times 2 \times 1}}{2 \times 1 \times (\cancel{3 \times 2 \times 1})} = 10$$

Thus, 10 represents the number of ways the singer could chose to sing three of five new songs. Therefore, the joint combination equals $15 \times 10 = 150$.

In summary, the outcome of this joint combination is:

$$_nC_r \times {}_nC_r = \frac{n!}{r!(n-r)!} \times \frac{n!}{r!(n-r)!}$$

$$_6C_4 \times {}_5C_2 = \frac{6!}{4!(6-4)!} \times \frac{5!}{2!(5-2)!}$$

$$_6C_4 \times {}_5C_2 = \frac{6!}{4!(2)!} \times \frac{5!}{2!(3)!}$$

$$_6C_4 \times {}_5C_2 = \frac{6 \times 5 \times 4!}{4!\,(2)!} \times \frac{5 \times 4 \times 3!}{2!\,(3!)} = 15 \times 10 = 150$$

14. Reunion

Choice D
Overview: This problem appears rather complicated on the surface, but its solution is actually quite simple. We're essentially asking: "How many groups of two can we create from a group of eleven where order doesn't matter? Or specifically: "How many ways can we choose two people from eleven people where order doesn't matter?"

$$_nC_r = \frac{n!}{r!(n-r)!}$$

$$_{11}C_2 = \frac{11!}{2!(11-2)!} = \frac{11 \times 10 \times 9!}{2!\,(9!)} = 55$$

NOTE ☙ This problem would have resulted in the same answer had it asked: "How many teams of two can we create from 11 persons in order to stage a beach volley ball tournament?" The formation of teams is a classic example of combinations in that order doesn't matter. It doesn't matter which member of a team is chosen first or second, a team of two is simply a team.

15. Outcomes

Choice A
Overview: This problem exists to test permutation and combination theory at a grass roots level. A strong understanding of theory will eliminate the need to do any calculations.

Statement I:

True. $\quad _5P_3 > _5P_2$

$_5P_3 = 60$ and $_5P_2 = 20$. Order matters in permutations and the more items there are in a permutation the more possibilities there are.

Statement II:

False. $\quad _5C_3 > _5C_2$

$_5C_3 = 10$ and $_5C_2 = 10$. Strange as it may seem, the outcomes are equal! "Complements in combinations" result in the same number of outcomes. Complements occur when the two inside numbers equal the same outside number. Here $3 + 2 = 5$. Note that this phenomenon occurs only in combinations and not in permutations.

Statement III:

False. $\quad _5C_2 > _5P_2$

$_5C_2 = 10$ and $_5P_2 = 20$. Order matters in permutations and this creates more possibilities relative to combinations. Stated in the reverse, order doesn't matter in combinations and this results in fewer outcomes than permutations, all things being equal.

About the Author

Brandon Royal (CPA, MBA) is an award-winning writer whose educational authorship includes *The Little Green Math Book, The Little Blue Reasoning Book, The Little Red Writing Book,* and *The Little Gold Grammar Book.* During his tenure working in Hong Kong for US-based Kaplan Educational Centers—a Washington Post subsidiary and the largest test-preparation organization in the world—Brandon honed his theories of teaching and education and developed a set of key learning "principles" to help define the basics of writing, grammar, math, and reasoning.

A Canadian by birth and graduate of the University of Chicago's Booth School of Business, his interest in writing began after completing writing courses at Harvard University. Since then he has authored a dozen books and reviews of his books have appeared in *Time Asia* magazine, *Publishers Weekly, Library Journal of America, Midwest Book Review, The Asian Review of Books, Choice Reviews Online, Asia Times Online,* and About.com.

Brandon is a five-time winner of the International Book Awards, a seven-time gold medalist at the President's Book Awards, as well as recipient of the "Educational Book of the Year" award as presented by the Book Publishers Association of Alberta. He has also been a winner or finalist at the Ben Franklin Book Awards, the Global eBook Awards, the Beverly Hills Book Awards, the IPPY Awards, the USA Book News "Best Book Awards," and the *Foreword* magazine Book of the Year Awards. He continues to write and publish in the belief that there will always be a place for books that inspire, enlighten, and enrich.

To contact the author:
E-mail: contact@brandonroyal.com
Web site: www.brandonroyal.com

Books by Brandon Royal

The Little Blue Reasoning Book:
  50 Powerful Principles for Clear and Effective Thinking

The Little Red Writing Book:
  20 Powerful Principles for Clear and Effective Writing

The Little Gold Grammar Book:
  40 Powerful Rules for Clear and Correct Writing

The Little Red Writing Book Deluxe Edition:
  Two Winning Books in One, Writing plus Grammar

The Little Green Math Book:
  30 Powerful Principles for Building Math and Numeracy Skills

The Little Purple Probability Book:
  Master the Thinking Skills to Succeed in Basic Probability

Ace the GMAT:
  Master the GMAT in 40 Days

Getting into Business School:
  100 Proven Admissions Strategies to Get You Accepted at the
  MBA Program of Your Choice

Dancing for Your Life:
  The True Story of Maria de la Torre and Her Secret Life
  in a Hong Kong Go-Go Bar

The Map Maker:
  An Illustrated Short Story About How Each of Us Sees the
  World Differently and Why Objectivity is Just an Illusion

Paradise Island:
  A Dreamer's Guidebook on How to Survive Paradise and
  Triumph over Human Nature

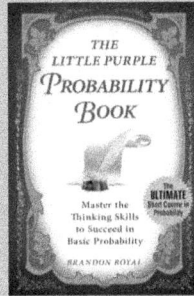

www.ingramcontent.com/pod-product-compliance
Lightning Source LLC
Chambersburg PA
CBHW060628030426
42337CB00018B/3259